Down Comes The

RAIN

by Franklyn M. Branley
illustrated by James Graham Hale

HarperCollins Publishers

Special thanks to Don W. Hen for his expert advice.

The illustrations in this book were done in pen-and-ink with watercolor washes.

The *Let's-Read-and-Find-Out Science* book series was originated by Dr. Franklyn M. Branley, Astronomer Emeritus and former Chairman of the American Museum–Hayden Planetarium, and was formerly co-edited by him and Dr. Roma Gans, Professor Emeritus of Childhood Education, Teachers College, Columbia University. Text and illustrations for each of the books in the series are checked for accuracy by an expert in the relevant field. For more information about Let's-Read-and-Find-Out Science books, write to HarperCollins Children's Books, 10 East 53rd Street, New York, NY 10022.

HarperCollins®, ☂®, and Let's Read-and-Find-Out Science® are trademarks of HarperCollins Publishers Inc.

Library of Congress Cataloging-in-Publication Data
Branley, Franklyn Mansfield, date
    Down comes the rain / by Franklyn M. Branley ; illustrated by James Graham Hale.
        p.   cm. — (Let's-read-and-find-out science. Stage 2)
    ISBN 0-06-025338-X.(lib. bdg.). — ISBN 0-06-025334-7 —ISBN 0-06-445166-6 (pbk.)
    1. Rain and rainfall—Juvenile literature.   2. Clouds—Juvenile literature.   [1. Rain and rainfall.
2. Clouds.]   I. Hale, James Graham, ill.   II. Title.   III. Series.
QC924.7.B694   1997   96-3519
551.57—dc20   CIP
                                                                                                              AC

Typography by Elynn Cohen
6   7   8   9   10

Newly Illustrated Edition, 1997

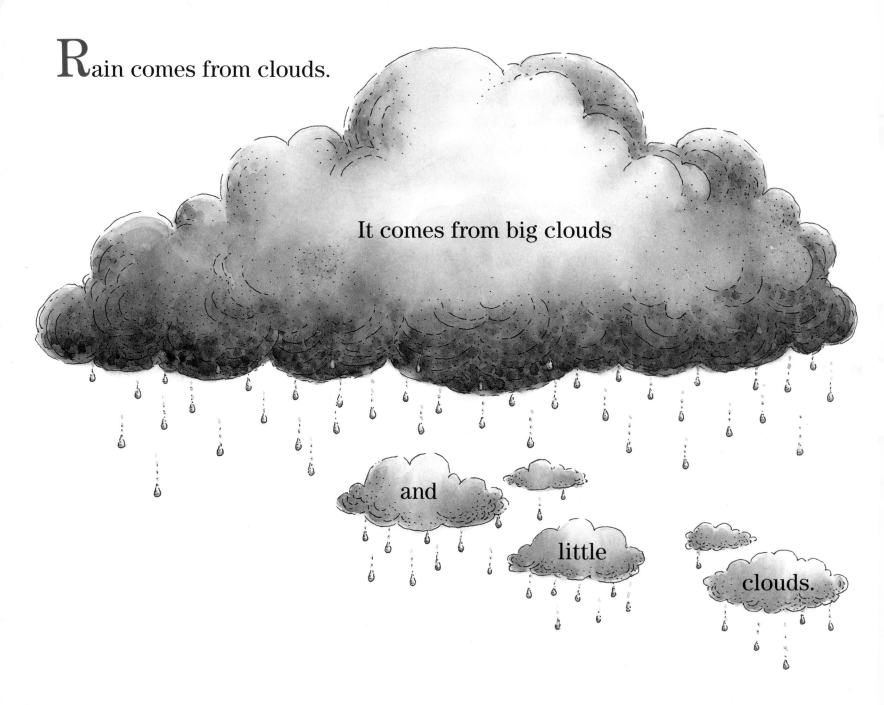

Rain comes from clouds.

It comes from big clouds

and little clouds.

It comes from black clouds, white clouds,

and gray clouds.

All clouds—big ones and little ones, gray ones and white ones—are made of billions of tiny drops of water. The drops are called droplets, because they are so small.

If this is the size of a drop of water,

a droplet would be just a tiny speck, even smaller than this one.

Water droplets come from water vapor. Water vapor is a gas.

Water vapor is made when water evaporates. That means the water changes from a liquid to a gas.

In the morning, put a teaspoon of water in a saucer...

By that night, it may have EVAPORATED into the air!

When you breathe out, you put water vapor into the air. Usually, you cannot see the water vapor. But it is there.

When cows, horses, dogs, and cats breathe out, they put water vapor into the air, too. On a cold day, the water vapor changes to droplets and makes little clouds that you can see.

You can make water vapor change to water.

Put a lot of ice into a glass of water.

As the glass gets colder, the outside of the glass gets WET.

Sometimes the glass stays dry. That means there is not much water vapor in the air.

The air holds the water vapor. Breezes carry it from one place to another.

Much of the vapor moves up and away from the earth.

17

A single droplet is so small you cannot see it. But you can see a cloud. That's because there are millions and millions and millions of water droplets in a cloud.

Sometimes there are only a few small raindrops
that fall slowly. It is drizzling.

Sometimes there are lots of big drops that fall very fast. Now it is pouring.

Sometimes the drops in clouds freeze. These raindrops become ice drops. This can happen even on a hot summer day.

Some clouds may be higher than most airplanes ever go.

The higher the clouds, the colder they are.

That's because the clouds and water droplets are high above the earth.

Many clouds are so high that it is freezing cold.

In these high, cold clouds, water vapor changes to droplets, and the droplets change to drops. The drops freeze into ice.

As they are carried upward, more water collects on the tiny bits of ice. When that water freezes, the drops of ice have another layer on them.

air currents

The ice drops are now heavier, so once more they fall toward the earth. But air moving away from the earth may carry the ice drops upward again. Higher and higher they go, and another layer of ice freezes onto them.

EA

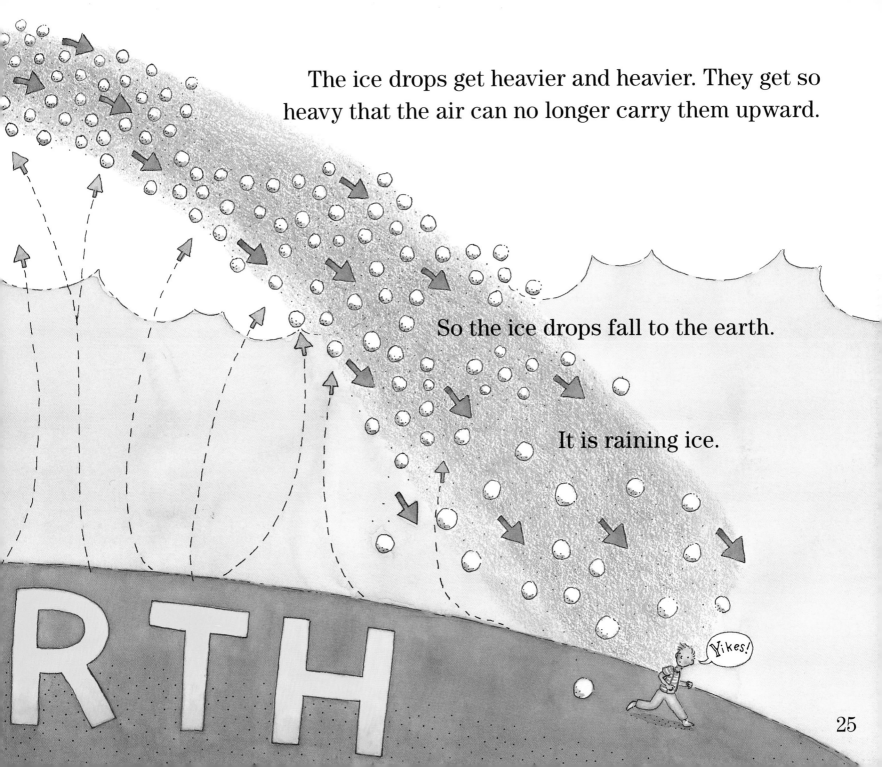

The ice drops get heavier and heavier. They get so heavy that the air can no longer carry them upward.

So the ice drops fall to the earth.

It is raining ice.

Yikes!

RTH

When the water vapor cools, it condenses.
The water vapor changes to water droplets,
and all together the droplets make clouds.

Water droplets join together to make water
drops. The drops fall to the earth from the clouds.

Once more it is raining.